ABCs of Disney's Zoo

ABCs of Disney's Zoo

by
JJ Waya

SmilingEagle Press Book

ISBN: 978-1-969767-03-6

Published by SmilingEagle
An imprint of Little Studio Publishing
For information:
SmilingEagle Press
www.smilingeagle.com

Printed in the United States of America

ABCs of Disney's Zoo

Introduction to ABCs of Disney's Zoo

Welcome to Disney's Animal Kingdom, the most amazing zoo in the world!

This book is designed to take you on a wild alphabet adventure through all the incredible animals that live at Disney's Animal Kingdom. We're going to jump right into the fun, starting with A is for Anteater and ending with Z is for Zebra. Each letter introduces you to a fascinating animal friend, along with silly facts that will make you laugh and specific locations where you can find them during your visit.

Whether you're preparing for a trip to Disney's Animal Kingdom, reminiscing about a recent visit, or simply love learning about animals from around the world, this ABC guide will entertain and educate. Each animal entry includes where to find them in the park, fun facts about their behaviors and characteristics, and just enough silliness to keep young readers giggling.

At the end of the book, parents and teachers will find a comprehensive educational guide section. This resource provides background information about Disney's Animal Kingdom's history and conservation efforts, discussion questions organized by age and learning level, activity suggestions, and tips for extending the learning experience both at home and in the classroom. The educational guide transforms this simple alphabet book into a complete learning tool that supports literacy development, science education, geography skills, and conservation awareness.

So buckle up for a wild ride through the alphabet! It's time to meet some amazing animal friends and discover why Disney's Animal Kingdom is truly the most magical place on Earth for animal lovers.

Are you ready? Let's start our ABC adventure!

A is for ANTEATER

Annie and Callie are always asleep! These adorable anteaters spend all day snoozing with their bushy tails covering them like fuzzy blankets. When they DO wake up, they can gobble 30,000 bugs for dinner. Yum!

Where to find them: The Oasis - left main trail.

Why Anteater over Aardvark? Because Willie the Aardvark is underground digging holes, while Annie and Callie are right there napping for everyone to see!

B is for BABIRUSA PIG

Betty the Babirusa pig has tusks that look like deer antlers! Her name means "pig deer" which is pretty confusing. Is she a pig? Is she a deer? She's just Betty being fabulous with her fancy curved tusks!

Where to find them: The Oasis - right trail, across from the muntjac deer

Your notes about this animal:

C is for CATTLE – ANKOLE, I MEAN

Ankole Cattle

Meet the Ankole-Watusi cattle, nature's walking hat rack with an identity crisis! These magnificent beasts sport horns so impressively wide they could probably get their own zip code, stretching up to 8 feet from tip to tip – which is convenient since they need the extra space to store all their complicated names (Ankole, Watusi, Inkuku... apparently they couldn't decide on just one). Weighing in at up to 1,600 pounds but blessed with hollow horns that won't give them a neck ache, these cattle are basically the inflatable pool floaties of the bovine world.

Where to find them: Kilimanjaro Safari Ride

D is for DUIKER (Yellow-backed)

The Yellow-backed Duiker is the newest kid on the block at Animal Kingdom! They're forest antelopes who prefer to live alone because they're basically the introverts of the antelope world.

Where to find them: Gorilla Falls Trail - sharing the exhibit with the Okapi

E is for ELEPHANT

These gentle giants are HUGE! A male elephant can weigh 7,000 pounds and drink 50 gallons of water every day. Their ears work like built-in fans to keep them cool. Talk about having big ears!

Where to find them: Kilimanjaro Safari - you'll see them twice during the ride!

F is for FRUIT BATS

The flying foxes in the bat house don't actually fly like foxes (because foxes don't fly, silly!). They're fruit bats who hang upside down all day. Maybe they think the world looks better upside down?

Where to find them: Maharajah Jungle Trek - in the special bat house you walk through

G is for GIRAFFE

The tallest animals at Disney! Their hearts weigh 25 pounds and their purple tongues are 20 inches long. Perfect for reaching the highest leaves and giving the biggest, slobberiest kisses!

Where to find them: Kilimanjaro Safari Savanna, also at Animal Kingdom Lodge

H is for HIPPO

Hippos look like they're always smiling, but don't be fooled - they're actually pretty grumpy! They spend all day in the water because they don't like getting sunburned. Smart thinking, hippos!

Where to find them: Kilimanjaro Safari - in the big pond, also at the fish tank on Gorilla Falls Trail

I is for IGUANA (Rhinoceros)

This spiky fellow loves to sunbathe on rocks like he's at the beach! The Rhinoceros Iguana has bumps on his nose that make him look tough, but he's really just a big vegetarian who loves flowers.

Where to find them: The Oasis - on the interior trail, sunning on a rock on your left

J is for JUMPING KANGAROOS

Red kangaroos can jump 25 feet in one hop! That's like jumping over a school bus. They use their tails like a fifth leg, which is pretty handy when you're a hopping expert!

Where to find them: Discovery Island Trails - best viewed from inside the trail near the Tree of Life

K is for KOMODO DRAGON

Meet Tiamat, Deidara, and little Khalessi! These dragons don't breathe fire, but they do have venomous bites. They're basically real-life dragons who forgot how to fly!

Where to find them: Maharajah Jungle Trek (the big ones) and Rafiki's Planet Watch (little Khalessi)

L is for LEMUR

Ring-tailed lemurs have tails longer than their bodies with fancy black and white stripes! They sit in yoga poses every morning to warm up in the sun. Very zen, lemurs!

Where to find them:
Discovery Island Trails -
near the Tree of Life, three viewing areas

M is for MEERKAT

These little guys are the neighborhood watch of the animal kingdom! One meerkat always stands guard while the others play. They're like furry security guards who work for free snacks!

Where to find them: Gorilla Falls Trail - between the bird aviary and the gorilla exhibit

N is for NAKED MOLE-RAT

The wrinkliest, most unusual animals at Disney! They live underground and look like tiny, hairless hot dogs with teeth. Beauty is in the eye of the beholder, right? **Where to find them:** Gorilla Falls Trail - in the dark Research Center building, look for the glass exhibit

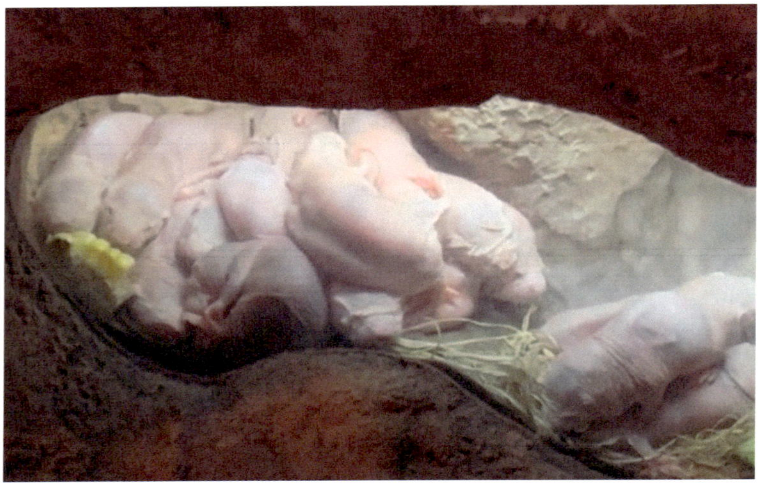

O is for OKAPI

The "forest giraffe" looks like someone couldn't decide between a giraffe and a zebra, so they made both! They have zebra stripes on their legs and a giraffe's long purple tongue.

Where to find them: Gorilla Falls Trail, Kilimanjaro Safari, and Animal Kingdom Lodge

P is for PORCUPINE

Peri the porcupine is basically a walking pin cushion! She has thousands of quills, and when she had baby Shelley, Disney keepers were very, very careful during cuddle time!

Where to find them: Discovery Island Trails - right side near the Galapagos tortoise (when not backstage) – or on Episode 6 of Disney+ Magic of Disney's Animal Kingdom.

Q is for QUICK SERVICE

When you get hungry looking at all these animals, Disney has quick service restaurants everywhere! Try the Dole Whip at Tamu Tamu - it's like eating sunshine in a cup! Or a vegan delight so the animals don't get frightened!

Where to find them: All over the park! Tamu Tamu (Africa), Isle of Java (Discovery Island), Yak & Yeti Cafe (Asia)

R is for RHINOCEROS

Disney has both black and white rhinos, but here's the funny part - they're both gray! The "white" rhino got its name from the word "wide" because of its wide mouth. Talk about a naming mix-up!

Where to find them: Kilimanjaro Safari - black rhinos come first, then white rhinos later in the ride

S is for SPRINGBOK

These antelopes can jump straight up 6 feet with all four legs stiff, like they're bouncing on invisible pogo sticks! The South African rugby team is named after them. Boing!

Where to find them: Kilimanjaro Safari Savanna

T is for TIGER

Sumatran tigers have more stripes than other tigers, like they're wearing extra-fancy pajamas!

Sohni, Anala, Conrad and baby Bakso are the stars of the Maharajah Jungle Trek.

U is for UNUSUAL WALLABY

Tammar wallabies are only 18 inches tall - they're like kangaroo puppies that never grew up! They can drink seawater, which would make us very sick but makes them very happy.

V is for VERY BIG WATER BUFFALO

Asian Water Buffalo ladies named after the Golden Girls live behind the tigers! Blanche, Dorothy, and Rose love to splash in their pond because buffalo don't have air conditioning!

W is for WILDEBEEST

The wildebeest is basically what happens when nature throws a "design a mammal" party after everyone's had too much to drink. These perpetually confused-looking creatures appear to be assembled from spare parts: they've got the head of a buffalo who's seen too much drama, the body of a cow having an identity crisis, and legs that seem borrowed from a horse who clearly skipped leg day. Every year, millions of them participate in the Great Migration, which is essentially the world's largest game of "follow the leader!"

X is for X-RAYS

Just like you get X-rays at the doctor, Disney's animals get X-rays too! At the special animal hospital, veterinarians like Dr. Dan and Dr. Natalie take pictures of the inside of animals to make sure their bones and bodies are healthy.

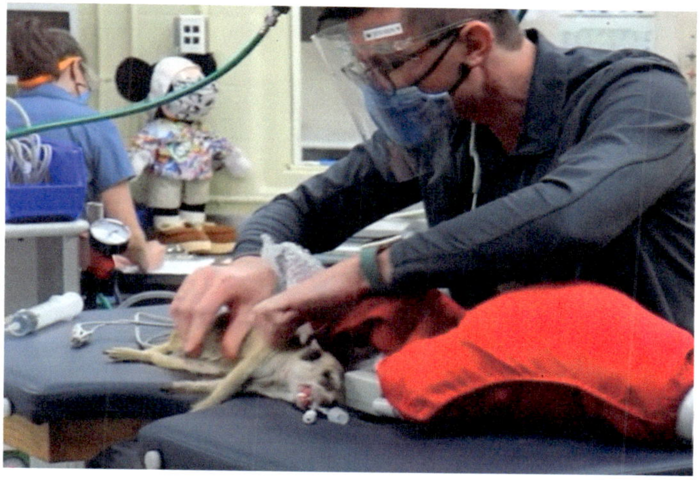

You can peek through the viewing window at Rafiki's Planet Watch and might see a meerkat getting a checkup or a tiger getting an exam!

Where to find them: Conservation Station at Rafiki's Planet Watch - look through the veterinary treatment room viewing window – take the train to get there!

Y is for YELLOW-BACKED DUIKER
Wait, we used Duiker for D! Let's try again...

Y is for YAWNING LIONS

Lions sleep 20 hours a day! They've got the best job ever - professional nappers who live on air-conditioned rocks. Living the dream!

Z is for ZEBRA

Three cheers for zebras! They're the perfect ending to our alphabet because they prove that life is better in stripes. Plus, they're excellent at standing out in a crowd!

Educational Guide

ABCs of Disney's Zoo is designed as an engaging early reader that combines alphabet learning with wildlife education and theme park exploration. This guide provides educators, homeschool parents, and caregivers with background information and discussion topics to maximize the learning experience.

Target Audience

Ages 4-8 (Pre-K through 2nd Grade)

- Early readers developing alphabet recognition and phonics skills
- Children beginning to read simple sentences independently
- Students learning about animal classifications and habitats
- Future or recent Disney visitors wanting to enhance their experience

Educational Settings:

- Homeschool curricula incorporating themes of wildlife, geography, and reading
- Elementary classrooms studying habitats, conservation, or theme-based learning
- After-school programs focused on literacy and science

- Summer reading programs with adventure themes
- Special education classrooms needing high-interest, accessible content

Disney's Animal Kingdom: It is More Than a Theme Park

Disney's Animal Kingdom is both a world-class theme park AND an accredited zoo, holding accreditation from both the Association of Zoos and Aquariums and the World Association of Zoos and Aquariums. As of 2019, it is the most-visited zoo in the world, making it a unique educational destination that combines entertainment with legitimate conservation efforts.

Key Facts About the Park:

- Covers 580 acres, making it the largest theme park in the world
- Home to approximately 2000 animals representing around 200 species
- Opened on Earth Day, April 22, 1998
- The park is dedicated and themed around natural environment and animal conservation
- Over 200 species of animals are living on the vast 540 acres
- Disney uses moats, water features and carefully disguised fences to keep animals in their habitats while making it appear they roam free

Special Animal Care Features:

- Animals have "backstage" barn shelters for nighttime and during dangerous Florida storms
- Some animals like anteaters Annie and Callie have "doubles" - one comes out in the morning, switches with another around noon
- Animals are frequently moved for medical checkups, habitat refurbishment, or breeding programs
- Disney hired staff from 69 zoos around the United States to care for the animals
- The park includes a veterinary treatment room with viewing windows for educational purposes

The History Behind the Magic

The Vision Begins (1989-1995) Disney began planning Animal Kingdom shortly after opening Disney-MGM Studios in 1989. The park was the brainchild of Imagineer Joe Rohde, who had previously designed the Adventurers Club at Pleasure Island. Michael Eisner, Disney's CEO, was concerned about Busch Gardens Tampa attracting potential Disney guests with its animal attractions and wanted to create something even more spectacular.

The Famous Tiger Presentation (1994) One of the most legendary moments in Disney history occurred when Joe Rohde brought a 400-pound Bengal tiger into the boardroom meeting with Disney CEO Michael

Eisner to demonstrate the excitement and power of live animals. This bold presentation convinced executives to greenlight the project.

From Concept to Construction (1995-1998)

- Originally announced as "Wild Animal Kingdom" in 1995 with an estimated cost of $600-$800 million
- The name was briefly "Wild Animal Kingdom" until concerns arose about confusion with Mutual of Omaha's "Wild Kingdom" TV show
- Michael Eisner came up with the final names "Animal Kingdom" and "DinoLand," saying "We have a Magic Kingdom so why not an Animal Kingdom?"
- DinoLand will be replaced with a new land in 2027, Tropical Americas.

Incredible Construction Details The landscaping efforts included spreading four million cubic yards of dirt, planting 40,000 mature trees (a mix of real Savanna species and artificial Baobab trees), constructing 60 miles of underground utilities, waterways, and structures built by over 2,600 construction workers. Disney Imagineers collected seeds from 37 countries to be used for the plants and grasses in the park.

Authentic Cultural Details Many buildings contained thatched roofs assembled by Zulu workers from South Africa. About 1,500 hand-painted wooden horses were crafted in Bali under Disney supervision. This attention to cultural authenticity makes the park an excellent teaching tool for world cultures and geography.

The Tree of Life Icon The Animal Kingdom's landmark Tree of Life is a 145-foot-tall man-made baobab tree with 325 animal carvings. It took 10 artists and 3 Imagineers working full time over 18 months to create this centerpiece that serves as both art and educational tool.

Understanding Animal Care and Conservation

Species Survival Plans (SSP) Many animals at Disney's Animal Kingdom participate in Species Survival Plans overseen by the Association of Zoos and Aquariums. These programs ensure responsible breeding for endangered species. Examples include:

- Babirusa pigs (Betty and her daughter Kirana)
- Sumatran tigers (critically endangered with less than 1,000 left in the world)
- White rhinoceros breeding program
- Okapi conservation efforts

Real Conservation Impact

- In 2006, Animal Kingdom rhinos were transferred to Uganda's Ziwa animal sanctuary to help re-introduce white rhinos to the country
- The park has successfully bred numerous endangered species
- Animals born at the park often participate in global conservation efforts
- Educational programs teach visitors about wildlife protection

Educational Benefits

Literacy Development:

- **Phonemic Awareness**: Each letter is clearly associated with animal names and sounds
- **Vocabulary Building**: Rich descriptive language introduces new words in context
- **Reading Comprehension**: Short, engaging paragraphs with humor help maintain attention
- **Cross-curricular Connection**: Links reading with science, geography, and social studies

Science Learning:

- **Animal Classification**: Introduction to mammals, reptiles, and different animal families
- **Habitat Recognition**: Animals are grouped by their geographic regions (Africa, Asia, etc.)
- **Conservation Awareness**: Subtle introduction to endangered species and protection efforts

- **Behavioral Science**: Fun facts about animal adaptations and behaviors
- **Anatomy and Physiology**: Learning about specialized features like giraffe hearts, elephant ears, and anteater tongues

Geography and Culture:

- **World Awareness**: Animals from different continents help children understand global diversity
- **Cultural Appreciation**: References to different countries and regions where animals originate
- **Map Skills**: Park locations can be used to teach basic navigation and spatial relationships
- **Architecture and Design**: Discussion of how Disney recreated authentic cultural elements

Discussion Topics and Activities

Pre-Reading Activities:

1. **Animal Prediction Game**: Show children the alphabet list and have them guess what animal might represent each letter
2. **World Map Exploration**: Point out Africa, Asia, and South America on a world map
3. **Zoo vs. Theme Park Discussion**: Talk about how Animal Kingdom is both entertainment and education

4. **Conservation Introduction**: Discuss why it's important to protect animals and their habitats

During Reading Discussions:

Animal Adaptations:

- Why do anteaters sleep during the day but hunt at night?
- How do elephant ears work like air conditioners?
- Why do hippos spend most of their time in water?
- How do giraffe hearts pump blood all the way to their heads?
- What advantages do porcupine quills provide?

Animal Behaviors:

- How does the meerkat "neighborhood watch" system work?
- Why do lemurs sit in yoga poses in the morning?
- How do tigers use their stripes for camouflage?
- Why do lions sleep 20 hours a day?
- How do wallabies conserve energy while hopping?

Conservation Questions:

- Why are Sumatran tigers considered critically endangered?
- How do zoos help save species from extinction?
- What can we do to protect animal habitats?
- Why is it important to have breeding programs for endangered animals?

Cultural Connections:

- How did different cultures traditionally interact with these animals?
- What can we learn from how Disney recreated authentic architecture?
- Why did Disney bring workers from different countries to build authentic structures?

Post-Reading Activities:

1. **Animal Habitat Sorting**: Group animals by their natural habitats (forest, savanna, water, etc.)
2. **Conservation Action Plan**: Discuss ways children can help protect wildlife
3. **Disney vs. Real Life**: Compare Disney's animal care with how animals live in the wild
4. **Plan a Pretend Visit**: Use the location guide to plan which animals to see first at Animal Kingdom
5. **Scientific Name Practice**: Learn to pronounce and understand scientific naming conventions

STEM Extensions:

- **Math**: Count stripes, spots, or other animal features; calculate animal sizes and weights
- **Science**: Research what different animals eat and create food webs; study animal adaptations
- **Technology**: Look up videos of these animals in their natural habitats; explore virtual zoo tours
- **Engineering**: Design an animal enclosure that keeps animals safe and happy while allowing visitor viewing

Writing Prompts:

- "If I could be any animal from the book, I would be... because..."
- "My favorite thing about Animal Kingdom would be..."
- "I think the most interesting animal fact was..."
- "If I worked at Animal Kingdom, I would help take care of..."
- "Animals are important because..."
- "To help protect endangered animals, I could..."

Assessment Ideas

Informal Assessment:

- Have students point out animals they remember while reading
- Ask children to retell funny facts about their favorite animals
- Listen for use of new vocabulary words in conversation
- Observe engagement level and questions during discussion
- Note students' ability to connect animals to their habitats

Formal Assessment Options:

- **Alphabet Recognition**: Point to letters and have students name the corresponding animal

- **Comprehension Check**: Ask simple questions about animal characteristics and habitats
- **Sequencing Activity**: Put animal cards in alphabetical order
- **Drawing Assessment**: Have students draw their favorite animal with one fact they learned
- **Geography Skills**: Have students place animals on a world map showing where they're from
- **Conservation Understanding**: Ask students to explain why certain animals need protection

Cross-Curricular Connections:

- **Social Studies**: Research the countries where featured animals live
- **Art**: Create animal drawings or sculptures inspired by the book
- **Music**: Learn songs about animals or conservation
- **PE**: Move like different animals (hop like kangaroos, walk like elephants)

Safety and Sensitivity Notes

Real Animal Education: While the book is fun and entertaining, it's important to remind children that these are real animals that deserve respect and protection. Disney's Animal Kingdom has special designs and provisions incorporated throughout the park to protect the animals' welfare.

Conservation Message: The book subtly introduces conservation concepts without being heavy-handed. Many of the animals mentioned are part of Species Survival Plans, showing how zoos and theme parks can contribute to wildlife protection. This provides opportunities to discuss environmental stewardship age-appropriately.

Cultural Sensitivity: The park's design includes authentic cultural elements from Africa, Asia, and other regions. This provides opportunities to discuss cultural diversity and respect for different traditions around the world, while acknowledging that Disney's representations are interpretations designed for education and entertainment.

Realistic Expectations: Help children understand that animals at theme parks and zoos have different lives than animals in the wild. Discuss both the benefits (medical care, protection, education) and limitations (restricted space, artificial environments) of animal care in captivity.

Additional Resources

- National Geographic Kids books about animals and conservation
- World map or globe for geographic reference
- Photos or videos of Disney's Animal Kingdom attractions
- Other alphabet books featuring animals for comparison
- Simple animal fact books for emerging readers
- Field guides to local wildlife to connect global and local conservation
- Another book published by SmilingEagle Press, **A Walking Tour of Disney's Animal Kingdom** by Sandi Jerome and her YouTube channel - Sandi Jerome with videos for most of the animals.

Digital References

- Disney's "Magic of Disney's Animal Kingdom" series on Disney+ (with adult supervision for age-appropriate content)
- National Geographic Kids website animal section
- Virtual zoo tours from accredited zoos worldwide
- Simple animal webcams from reputable zoos and wildlife preserves
- Interactive websites about animal conservation

Field Trip Connections

- Local zoos and aquariums
- Wildlife rehabilitation centers
- Natural History Museums
- State or national parks with wildlife viewing
- Environmental education centers

Extension Reading

- Books about specific animals featured in the ABC guide
- Stories about veterinarians and animal caretakers

- Picture books about conservation heroes
- Biographies of wildlife researchers and conservationists

This educational guide transforms a simple alphabet book into a comprehensive learning experience that builds literacy skills while fostering curiosity about the natural world, conservation, and global cultures. The combination of humor, factual information, and practical application makes it an ideal resource for diverse learning environments, from traditional classrooms to homeschool settings to informal family learning experiences.

About the Author

JJ Waya is an emerging children's author passionate about combining education with entertainment. With a background in technical writing, JJ brings a unique perspective to children's literature that makes learning fun and accessible.

As a lifelong Disney enthusiast and frequent visitor to theme parks around the world, JJ has spent countless hours observing and researching the animals that call these special places home.

JJ holds a technical degree and has worked as computer consultant and. This technical background brings a systematic, research-based approach to children's educational content, ensuring that every fact is accurate and age-appropriate.

When not writing, JJ enjoys sketching wildlife, visiting zoos and aquariums, and exploring national parks. A strong advocate for conservation education, JJ believes that inspiring children to love and understand animals is the first step in creating the next generation of environmental stewards.

This is JJ's first published children's book, with plans for additional educational series covering birds, marine life, and plant ecosystems in theme park and zoo settings.

JJ currently resides in Florida, where year-round access to Disney's Animal Kingdom provides ongoing inspiration for wildlife education and conservation storytelling.

About the Author

JJ Waya is an emerging children's author passionate about combining education with entertainment. With a background in technical writing, JJ brings a unique perspective to children's literature that makes learning fun and accessible.

As a lifelong Disney enthusiast and frequent visitor to theme parks around the world, JJ has spent countless hours observing and researching the animals that call these special places home.

JJ holds a technical degree and has worked as computer consultant and. This technical background brings a systematic, research-based approach to children's educational content, ensuring that every fact is accurate and age-appropriate.

When not writing, JJ enjoys sketching wildlife, visiting zoos and aquariums, and exploring national parks. A strong advocate for conservation education, JJ believes that inspiring children to love and understand animals is the first step in creating the next generation of environmental stewards.

This is JJ's first published children's book, with plans for additional educational series covering birds, marine life, and plant ecosystems in theme park and zoo settings.

JJ currently resides in Florida, where year-round access to Disney's Animal Kingdom provides ongoing inspiration for wildlife education and conservation storytelling.

Made in the USA
Monee, IL
05 November 2026

67382716R00033